# The Rough Journey

## Bootstraps and Beyond

Dorothy Tolliver

BK
ROYSTON
Publishing

BK Royston Publishing
P. O. Box 4321
Jeffersonville, IN 47131
502-802-5385
http://www.bkroystonpublishing.com
bkroystonpublishing@gmail.com



Cover Design: Elite Book Covers

ISBN-13: 978-1-959543-28-2

Printed in the United States of America

**Illustrated by Kenneth Wade Tolliver**
**Grandson**

# DEDICATION

I dedicate this book to the memory of my mom and dad. It is because of them I have great memories of farm life and learned that there is more to life than bright lights in the big city.

These are memories of lives well lived during a time of segregation and perseverance.

# ACKNOWLEDGMENTS

I thank my Lord and Savior Jesus Christ for giving me the opportunity to leave behind memories to my children. This is something that I have wanted to do for a long time.

To Janet, my daughter: she loves to tell stories about her and Granddaddy getting into a big discussion and Grandmother passing along good food, sometimes with directions how to prepare it.

To Kenneth, my son: I wish he had more time with Grandaddy, so he could have learned more about farm life.

# Table of Contents

# INTRODUCTION

The year is nineteen fifty-two. Farm life, sharecroppers, segregation, and hard times. I didn't know I was considered a poor Negro, until a larger population of the white community told my dad.

I was fine with a place to stay, food to eat, friends to associate with, and the support of a loving family.

I want to leave behind a written word of my family life: the good, the bad, and the ugly.

I hope to reach families and tell them life can be hard, but to never give up. There is a brighter day coming.

You can make a life for yourself, if you have determination and a plan to look forward to.

# BOOTSTRAPS AND BEYOND

The forty acres and a mule were no more. Years of the mule and wagon gave way to the motorized equipment down on the farm.

Imagine, if you can, a Negro farmer from the age of nineteen, finally starting his own farm at the age of forty-something after working on someone else's farm for years. The farm was off a country road, overgrown with underbrush, road in rutted conditions due to lack of up-keep.

Mom, Dad, Sister, Grandmother, and I moved from the old farmhouse, where I was born, to a place we could call our own. With no boys in the family, it was the job for us girls and Mom to help Dad with the everyday chores.

Days were long in summer. You could start milking the cows at six in the morning, come back to the house for breakfast, and start somewhere on the farm in a field that either had a garden or crop that needed to be tended.

At day's end when we left the field to go to the house, sometimes we would take a shortcut across some of the fields. We would forget that some of the sheep would chase you if they saw you. There's nothing like going over a barbed wire fence with a dress on and getting caught in the wire with the sheep about three steps behind. If you made it over without getting a head butt from the sheep, your clothing was usually not as lucky.

Sheep shearing was another long, hot smelly job. Daddy and Mr. Taylor, the farm owner, would shear sheep for many days before the herd was finished. They sheared the sheep to keep the sheep cool in summer. At the end of each day, Daddy and Mr. Taylor would come home oily and smelly from the job. They bundled the wool and took it to Russellville, Kentucky, to market. Daddy was paid for helping with the job, but I never knew how much.

Bath time after this job was a welcome time. A good hot bath, in the number two wash tub, with a bar of the lye soap that Mom had made from lye, bacon drippings, and cold water, made Daddy look and smell like a different person.

After Daddy milked the cows and strained, or filtered, the milk into a ten-gallon milk can, there was a milk truck that came each day to all the farms that had milk to sell at the creamery. This was extra money, as was selling eggs at the country market to buy extra things like sugar, flour, pepper and salt to add to the food supply to keep the home in operation year-round.

While the milk was in the can for a period, the cream (fat) would rise to the top and Mom would remove (skim) it from the milk container, put it in half-gallon jars, and my sister and I would rock the jar back and forth until it became fresh butter. It had the freshest, most delicious taste ever, and on those mile-high cat head biscuits, at breakfast, it was delightful. Now it's time to go to the field.

The sun beamed down in the middle of summer, and the only relief was a shade tree at the end of the row of tobacco and maybe a container of ice water. Up and down until noon. Soon it was time for lunch. Sometimes we went to the house to eat; other times we took sandwiches to the field with us. Then and only then did you leave the field until day's end. Sometimes for miles around you could hear other men, women, and children in adjourning fields whistling or singing and children playing. Dad would always call out to us to be careful, saying, "Don't step on the tobacco." At day's end it was time to milk the cows again and check on the other livestock.

Grandmother would have helped Mom with dinner and supper. The breakfast meal would be as large as the other ones. Most everything was raised on the farm. The meat and vegetables were in great quantity, especially in early spring and summer, but still what was in the garden had to be harvested for winter. This made days much longer after being tired from being in the field all day, working the tobacco crops, and getting the vegetables canned for winter.

Canning the vegetables was a long, hot day's work with no air conditioning, only the open windows that let in a cool breeze every now and then, which was not often felt, because of the wood stove that was used to prepare the food for canning. Most mornings before going to the field found the family in the garden gathering the vegetables before the dew's moisture was off the vine. We all snapped the green beans. It was like a quilting bee without the pretty fabric. Squash, beets, peas, and many other vegetables required our attention, also.

The dreaded days of hard work were no longer remembered when winter came, and you didn't need to go to the neighborhood grocery store or drive ten miles to the city to buy most of the ingredients for a meal. Most days there wasn't money in the till to do so, anyway. So, we made do with what we had. Even when Daddy went to the grocery, he only bought flour, meal, and sugar; at other times he bought molasses, but it was bought in large quantity also. Now we could go for a few more days or weeks before buying more. But there wasn't a need to if Mom and Grandmother had made blackberry preserves, peach preserves, or apple butter during the season when the fruit was plentiful. As you ate the preserves and jellies sometimes you would remember the chigger bites you received when you went to pick the berries. It didn't matter how much coal oil, kerosene, or long sleeves you put on, they always found a way to get around your waist or under your arms to cause an itch that was horrible and needed more medicine to get rid of the pain.

My sister and I learned very early in life how to harvest vegetables, wash the jars for them, and put the vegetables in for canning. Not only did we harvest the vegetables, but in the winter Dad would also slaughter pigs for us to have meat for the winter months. It became our job to help carry large containers outside for the adults to put the fresh meat in before it was either cooked and put in jars or to have samples of the sausage to taste, to adjust the seasoning, before it was ready to be packaged in the muslin sack and hung in the smokehouse to begin the smoking process. This method cured and prevented them from spoilage. It could be eaten when you needed them or when it was time for a big family holiday gathering.

## Grandmother's Stories

Let me take you back to the time of my birth at the old farmhouse that I left when I was ten years old. Grandmother would tell stories of what happened in days gone by. She was born in eighteen ninety-nine. She and Daddy lived on the farm where Daddy sharecropped from the age of nineteen until he was about forty years old. He was the head of the house, with no father. Grandmother after a few years became a midwife. She helped one young doctor deliver his first baby, after he started his practice, at a home near Russellville, Kentucky. She stayed with the mother for a few weeks until they could manage on their own again. This was the practice back in the days of old.

In her spare time, she would take me, my sister, the boys on the farm, and some of the neighbor children fishing. She always carried a chair to sit in, her basket, an old corn cob pipe, tobacco, and a can filled with fishing worms. She never had any of the new-fangled rod and reels, just a cane pole with line, hook and sinker, and a weight. You could see the pleasure on her face as she puffed on her pipe and told us kids to be quiet before we scared the fish away.

Mrs. Barnett, grandmother of two of our friends, always put the finishing touches to many of our outings by preparing a fresh peach cobbler with a white sauce. After we finished a meal of fried chicken and other side dishes, it wasn't any trouble to keep us quiet. Late evening would find us walking back to our various homes with our catch of the day, hopefully to return to the fishing hole in a few days for more fun.

Grandmother invented ways to keep us occupied. She was a great quilter, and she would cut out pieces of material for us to sew, just to keep us from her stash or jumping on the springs of the old iron bed. I managed to finish one quilt top, and the eldest son of the farm owner did also. Mine wasn't pretty, but his was so perfect, after his was finished, his mom had it quilted and entered it in the county fair. A blue ribbon, first prize, was his for the taking. Now his wife has it on display in her home. For you see, he grew up, worked on the farm, and as a young man was killed in a tractor accident a few years after his second child was born.

My mom made string quilts. She put strips of material on squares of newspaper or a brown paper bag to get a square measure. These were made from old dresses and other materials, mostly materials from feed sacks that had a pretty pattern on them, or maybe from old suits that were tacked and quilted, made into a quilt for winter, and kept you very warm.

## Daddy Moved..

When Daddy moved from the farm, I was ten years old. It seemed we would never see our friends again. Many years have gone by. My sister and I, as well as Mrs. Barnett's grandchildren, and the brother of our friend who was killed in the farm accident, have grown up, married, and have children and some grandchildren of their own, remain friends today.

My dad had many hard years on the farm because he started farming with mules and hand plows, and worked from sunup until sundown to clear the land, put in new fencing around the property, and repair the rutted road that led home.

With no brothers to help clear the land and put out crops, or many other things that needed to be done to keep the farm going, Mom, my sister, and I became the hired hands, except we never got paid extra.

Fall weather would find us helping harvest the summer crops, finish canning vegetables from the garden for winter and help with the slaughtering of pigs to have meat for winter.

The old farmhouse we lived in wasn't a mansion, but we kept it clean and livable. Winter nights found the family gathered around the fireplace or the warm heater to keep us warm in the morning. Stories were told to us at bedtime, or so it seemed to me and my sister. After everyone was in bed and the nights grew longer, we kept warm under the many quilts that Grandmother and Mom made.

Early morning would find the fires burned low and Daddy getting up to put coal on the fire or even starting a new fire for the day. Fires were started from wood chips that my sister brought in the evening before. After the house was warm, Mom and Grandmother would get up and start the day. Daddy went to the barn to milk the cows and feed the pigs and chickens.

Breakfast was something to die for — sausage from the slaughtered swine that had been canned earlier in the season, eggs from the chicken house, homemade biscuits that were a mile high and so flaky they would fall apart as you ate. Oatmeal, which I didn't like but my sister thought was a delicacy, milk from the day before, and finally, coffee perked in the old aluminum coffee pot. It had the greatest aroma in the world. We children wanted to taste it, but Grandmother always got around giving us a drink by saying it would make us black, as if we weren't already. Breakfast over, dishes washed and another day in progress.

Mom and Grandmother started dinner, and Grandmother tended the many pots and pans on the wood stove. Mom gave us toys to keep us busy while she went to the barn to help Daddy strip tobacco. It had cured or "come in order" — a term used to say tobacco was ready to be put into different piles, tied together, and get ready to be sold at the tobacco market. The days were cold, and the only heat was a fire on the ground in front of the door to keep them warm. Mom and Daddy would work until Grandmother called for dinner, or they got to a “stopping point” and come back later to finish for the day. By day’s end, the cows needed milking again, the pigs and chickens fed, wood chips gathered for fire in the morning, and the cycle started again.

Before bedtime, my sister and I would sit in Mom’s or Grandmother’s lap and listen to them read to us or just tell us stories that kept us entertained until we fell asleep, and they put us to bed, or we crawled into bed and fell asleep under the warm quilts again. Sometimes if the night was really cold, and Mom thought the bed was not warm enough, she would warm a brick by the fire and wrap it in a towel at the foot of the bed to get us warm.

The winter months seem to last forever, and Thanksgiving would bring about the cooking of the fat hen from the chicken coop, cornbread dressing seasoned to perfection, and mashed potatoes made from the harvest in early fall that had been stored in the closet or in the barn under straw to keep them from freezing. Of course, you couldn't forget the green beans that tasted as fresh as the day they were picked from the vines. I can't remember the other vegetables because there were many, many more.

Desserts included at least two or three cakes and four pies. My grandmother's orange cake, a fresh coconut cake, and egg custard pies that had meringue that was a mile high and had been baked in the oven until it was a golden brown.

Company, visitors, was all the neighbors from miles around, including the minister from Sunday service. Leftovers were always better the next day, and believe you me there was plenty. An army could still be fed.

## The Fall and Christmas Time

The excitement of Christmas in the air was fun. The smell of the fresh cut cedar tree, in a bucket of rocks and water to keep it from shedding its needles off the branches, still brings back memories today. I remember the fresh fruit we could smell (even though Mom said, "It is your imagination") that had been hidden under the bed in a wooden box. In our stocking on Christmas morning, we found an orange, apple, mixed nuts, and a foot-long peppermint candy stick. We made the peppermint stick last as long as possible by breaking off small pieces at a time. We believed in Santa Claus and always left him cake and boiled custard. Christmas, like Thanksgiving, would find the house clean and plenty of food ready for company to come. Getting ready meant at least eight cakes, an old ham from the smokehouse that would be cooked in a lard can that either had all the lard used from it after the slaughter of the pigs, or one that was kept especially for the occasion. Again, all the vegetables from the summer canning and the potatoes from the barn, to be finished before it was time to plant again, or they would spoil before they could be eaten. There was enough cake left to last until the New Year, but sometimes Mom had to make another gallon of boiled custard.

We never took the Christmas tree down until after the first day of the year. This was supposed to be good luck. We always left our gifts under the tree until then only taking our toys out to play with and put them back at day's end.

The winter days were still long. Most of the tobacco had been stripped and taken to market.

At the end of summer with most of the vegetables prepared for winter, the last crops would be ready. Those would be peanuts, which we could roast on a cold winter night, as a snack before bed, or popcorn that had been harvested, shelled, and put into tin containers to stay fresh and dry until we were ready to pop it as we listened to the old battery-operated radio and tune to the Friday night frolic from Nashville, Tennessee. Finally, when the sweet potatoes were ready for baking, we would rub them in bacon dripping, wrap them in foil, place them in the old fireplace or stove oven until done, and put homemade butter and sugar on them to eat.

Fall was time for picking corn. Daddy had to pick many acres by hand, as my sister Josephine and I drove the mule and wagon down the field. If that wasn't tiring enough, when he finished, all the corn was put in the corn crib. Large shelves were used to scoop the corn from the wagon into the crib. This kept the corn dry to feed pigs and chickens if they didn't have special feed for them. If we shelled any of the corn, we kept the cobs to start fires in the wood stove or dump in the pig pen for the pigs to have a snack.

By now, all the walking around the farm has worn Daddy's shoes to the point of the soles pulling away from the leather. The year was coming to an end, and many times he made do by using pieces of wire to hold them together until he sold his tobacco and had enough money to buy another pair.

Many times, Mom patched his overalls to hold them together, also. When he bought his shoes, usually he bought new overalls and a jacket to complete the outfit and several pairs of gloves.

## Spring

Now it's time to think about spring. It will not be long before its time to prepare the soil for tobacco plants. Time to plow the fields and get ready to plant corn and garden.

Let's not forget spring cleaning. More work!!! We took the feather tick mattress (or maybe a straw one) off the bed and hung it on the clothesline to get fresh air after a long winter. Also, we did window washing and painting the chairs and anything else that looked worn to freshen them up for the summer. It's time for a new linoleum for all the floors and an oil cloth for the kitchen table. After many long days of moving furniture from side to side, the old farmhouse took on a fresh new smell and look.

By now it's warm enough for the chicken to have laid enough eggs to have extra eggs to put in a nest where a mother hen would think it was time to hatch baby chicks. Once they hatch and leave the nest, the grass in the chicken yard would be tall enough for them to eat. By now, the guineas, ducks, turkeys, and most anything that had feathers or fur had little ones running around on the farm. The cutest were the baby ducks and the baby calves.

One Sunday morning, I let my pet guinea out of his cage to eat grass before I went to church. He was a small white creature that thought he was as large as anything on the farm. The big red rooster had other thoughts, and the fight was on. The rooster spurred and killed my pet. I became upset and cried so much I had to lie down after I arrived at church. I had a funeral for him and buried him at the edge of the wood pile, in the back yard. I still visit his grave site when I go home.

The little runt, small piglets were brought from the mother pig to the house and put in a wooden crate and we fed them with a baby bottle until they were big enough to go back to the mother pig. We had two that stayed longer than usual. We named them by that time Ruddy and Squeeky. You could put Ruddy on a towel in your lap and scratch her back. As long as you did that, she would sleep the time away. Squeeky wasn't as loving. He was as ornery as the days were long, because when he was big enough to run around in the yard, he would almost make Mom fall by running up against her legs. The bad thing was he knew he was wrong because he would run away and look back to see if anyone would chase him.

Sunday afternoons, most of the time, would be the day the family would put the water containers in the truck and go to the creek to get water for wash day on Monday. The ford crossing was low enough to fill the large barrels without a lot of effort. Back at the house we filled the large iron kettle and got ready to start the fire early Monday and start the family wash. Grandmother and Mom took turns on the washboard getting things clean. I do believe the white clothes, sheets, etc., were white as snow and looked pretty hanging on the clothesline. The tough times came when it was time to wash Daddy's overalls. They were heavy and tough, but so were the quilts and blankets. After all the washing on Monday and getting them off the line, Tuesday would be the time to iron all that needed to be done. Everything that had been starched was sprinkled with water and put in a plastic bag. This would give them time to soften and be easier to iron. Grandmother ironed everything — sheets, pillowcases, towels, socks, underwear, and anything else that happened to be in the stack of clothes. The starch was homemade with flour and water. It really got the job done. The clothes would almost stand-alone but look so crisp when you put them on. The ironing was done with a flat smooth iron that had been heated over an open flame outside in summer or on the wood stove in winter since we had no electricity.

It's spring and time to plant the garden. For fertilizer, Daddy would clean out the chicken house floor and put the manure on the ground, plow it into the soil, and let it rest for a few days before planting the seeds, otherwise the seeds would die, because the soil would be too rich for the plant to survive. Once all this was done, the planting began — green beans, squash, beets, cabbage, and tomato plants. They were planted around the tobacco plant bed. The seeds were planted when Daddy did his tobacco seeds.

The never-ending time for work continued day in and day out. The garden is finished. In a few days it will be time to pull tobacco plants to set in the fields. As seedlings of the garden and the tobacco plants grew, the ground needed to be cleared of any weeds growing, so this was the time to get the hoe and clean between the rows, so everything could grow, and grow it did. The first vegetables to come from the garden were fresh and tasty after cooking. The green beans with a big cut of jowl bacon, from the smokehouse, made them delicious. Since the smoked meat was now getting scarce, after feeding the family all winter, in a few months you think of the slaughter again.

The tobacco has grown enough to have worms on the leaves as well as the extra growth known as "suckers," that needed to be taken off the big plant for it to finish its growth. Removing the worms wasn't fun. Most of the time you pulled them by hand and destroyed them, so they wouldn't get on the plant again. It might not have been hard if we only had a row or two of tobacco, but it was about five or seven acres. This had to be done several times before the tobacco was ready to cut, and unless it was a cloudy day, it was in the hottest time of the day.

## From the Old Farmhouse to the Farm

Preparing to move from the old farmhouse to the farm we would now call home was mixed with tears of sadness and joy. The Taylors wanted us to stay, but Daddy knew it was time to move on. With everything packed and many trips made, we were now at the homeplace. The house was larger than where we had been. Now we had electricity but still no inside plumbing, so for many years we had a two-seater outhouse at the rear of the chicken house. We had a water well at the back door, and we no longer had to carry water in cans from the house down the road or fill barrels from the creek to do the family wash, so this was great.

We were afraid for a while because we had been told the house was haunted. Maybe it was, but when we showed no fear and invited who or whatever opened the stairway door to come in and have meal with us, that all stopped, and life in our new dwelling began to be a place of pleasure. With furniture in place, kitchen pots and pans put away, things were coming together. It was winter when we moved, so we had a lot to look forward to as the seasons continued to change.

The road was not repaired yet, so we walked by lantern light from the traveled road at night by way of the access road to the house. We were scared, but Daddy always took the lead, and this became a way of life if we were away from home after dark.

Our new home was beautiful. It was built in what was known as the “gingerbread” style. The porch had all cut work near the top of the poles, and the windows covered a floor to ceiling position. There were two fireplaces, one in the living room, the other in the dining room. As it was, we did use wood and coal for heat. The years have gone by now, and our high school years are approaching. Integration is taking place, and the schools, buses, many other changes are happening, also. The high school years are proving to be interesting. Even though things are going well, everything is not as smooth as you hope for. There was always a few of the white population that was not receptive of you. I can recall the day I made a higher grade than one of my classmates, and of course everyone discussed test scores when the papers were returned. Before the discussion between me, and the other student, and teacher, my score was reduced by a point. It was hard for the student to believe I could be as good as they were. Therefore, the teacher decided to reduce my grade by one point to keep the other student happy.

I didn't get to attend prom night as part of the class. No blacks allowed. After graduation was over, I returned home with my parents. It did upset me for a while, then I realized I was the fortunate one. I had not offended anyone, except for the color of my skin, and that was their problem.

My sister did go to her prom three years later and had a marvelous time. Times had improved and more acceptances was prevalent.

In my junior year of high school, our beautiful farmhouse burned shortly after the

Christmas holidays. Grandmother was home, but she couldn't remember what happened. The family dog, Rattler, suffered smoke inhalation trying to keep her from returning inside the house. He passed away a few weeks later.

Mom, Dad, sister and I had gone to visit my other grandmother, Mama Idella. Mom said that she had a strange feeling something was wrong, but couldn't put her finger on why. Our neighbors tried to save the house but could not because the water source was at the back door. Returning home, we found only a chimney standing. It felt as if a family member had passed away.

We had cousins who lived in the Gordonsville community that we stayed with for a few days. The George Hardin family had a vacant house, which they allowed us to use until we could do better. About a week or two had gone by, and with donations we had enough furniture to start housekeeping again and enough clothing for all the family to have several changes. The neighborhood, Negro and white, came together to help us, and the sad memories began to fade. Life was beginning to be normal again.

## School

Olmstead High School issued my sister and me books to finish the school term. I'm in my senior year now, and it's time to order invitations, cap and gown, and class ring. My class ring cost about twenty dollars, but Daddy had to borrow the money since he was in the process of trying to rebuild the family home. The house is built, now it's time to move again. With everything almost in place again, life took on a new normal.

Graduation has taken place and a neighbor asked if I would consider working in the dietary department at the local hospital. This was a good move for me, since I didn't get to go to college.

In a couple of years my sister will graduate from Olmstead High School. She is talking about going to college. I felt there wasn't enough money for me to attend college, so I continued to work. I wasn't given any information when I graduated about scholarships or money I could get to attend. Graduation came for my sister, so now she is wanting to attend college. By the grace of God, somehow she went to college and graduated by working her way through her years at Kentucky State College, Frankfort, Kentucky, with a degree in home economics. While at Kentucky State, she met and married the love of her life, moved to Detroit, Michigan, taught home economics for a while, went to Wayne State University, Detroit, received her Master's degree and did counseling for one of the high schools and retired after thirty-three and a half years.

I was the oldest and the first to get married and leave home. My wedding wasn't fancy, but it was beautiful. It was held outside on the front porch. June twenty-seventh nineteen sixty-five. The lawn was crowded with neighbors and family from miles around. My sister's was held at the same location, June twenty-second nineteen sixty-eight. The announcement for both occasions invited family and friends to attend. Everyone told Daddy that they didn't care if they were invited or not, they would be in attendance, since they had known him since he was a young man of the community for many years.

I worked for five more years at the local hospital and was fortunate enough to learn my skills associated with many of the departments of the institution. Still some people were not receptive of the idea of a Negro overseeing a section, which made it difficult to enjoy a day at work. I made up my mind that the best I could do would be my epitaph. No matter how hard the chore, make it a thing of joy. Working in that position before I knew what was happening, I became the person that made most of the desserts for patients and employees. Pies and cakes I didn't mind, but cookies were not my thing. Maybe I wouldn't have dreaded it much If it had not been the chore of making enough cookies for the annual Christmas party, which was about three hundred people.

I married five years after graduation and moved to the community of Fairview, about seventeen miles from home. I always laughed and told Mom I didn't want to move far from home, because If I wanted to come home for a visit, I didn't want a long travel time.

Education was not the foundation for my dad, since he had to be the man of the house. With no father figure around, it was he and his mom. What he lacked in education was overcome by wisdom, watching and listening to how the world evolved. He was not afraid of working and could answer questions about farming, even though his education was short lived. You see, he only went to the third grade. His first-grade teacher was also the teacher for me and my sister. Our transportation to school for the first six years were with her. She passed away in February two thousand eight, at one hundred years of age.

## Retirement

It's June nineteenth nineteen ninety-nine, and Daddy had decided to sell his farm equipment and retire from farming. Instead of the mule and wagon, he now had tractors, trucks, combine, manure spreader and everything that made farming much easier. He decided to maintain the vegetable garden and drive to the city for clients he had previously served. They wanted nice clean fresh vegetables from the country.

From nineteen years of age until eighty-two, the Negro farmer endured prejudice, long days without help around the farm, had the attitude of "I can and I will." By now he was making enough from the once run-down farm to white people working for him and being paid good wages for a day's work.

Mom still cooked the big meals for everyone to enjoy. We often thought that is why they didn't mind working, because the kitchen was always a place where Negro and white sat at the same table together and enjoyed a good meal.

Mom wasn't in a hot kitchen anymore up at the "big house." She was in her own kitchen with electricity, running water, and a time schedule all her own.

# WORDS OF WISDOM FROM DADDY!!!!!

Work hard and save your money.

Never follow the wrong crowd.

Come home before midnight.

Stay out of trouble and jail.

That joker ain't worth a quarter.

The main thing is-------------------.

# THINGS MADE ON THE FARM FOR CLEANING AND A PART OF EVERYDAY MEALS

## LYE SOAP

1 can of lye (store bought)

1 gallon hot water

4 lbs. (pounds) clean bacon drippings or old lard, not in use

Mix lye and water, carefully mix in the melted grease, stirring well with a wooden spoon

Pour into an iron or enamel pan

Let it harden, cut into bath bar size

Good for washing clothes, taking a bath, and washing your hair

Make ahead and let it age for several weeks, the older the better

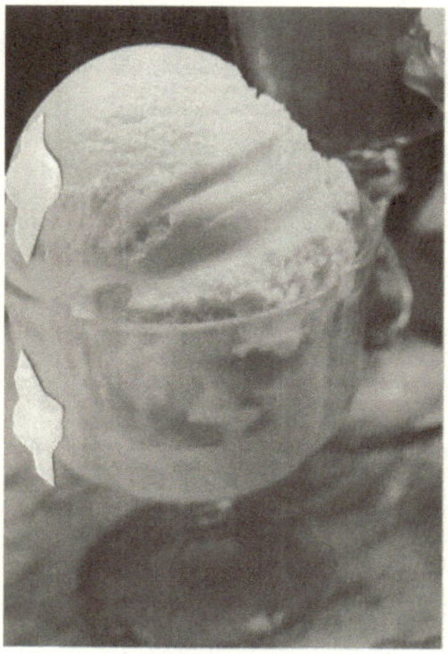

## VANILLA ICE CREAM

In summer our pleasure on Sunday was homemade vanilla ice cream. Mom would make the custard, put it in the hand cranked freezer, pack around it with ice and rock salt, and my sister and I would take turns cranking the handle until it was hard to turn, a sign that the custard was frozen and ready to eat. I believe that was the best ice cream ever. Sometimes we had cake to eat with it or what we called a “two egg puddin’.”

## CUSTARD

8 eggs beaten slightly

1 gallon of milk

1¼ cup sugar

Mix sugar and eggs until smooth, add milk until all mixture is smooth. Pour in a double boiler over medium heat. Stir often until the custard is thick and coats the spoon. Remove from heat. Add one teaspoon vanilla flavoring. Put in freezer with top on and place the crank on. Pack ice around the container and rock salt and the crank. Turn the crank until it no longer turns.

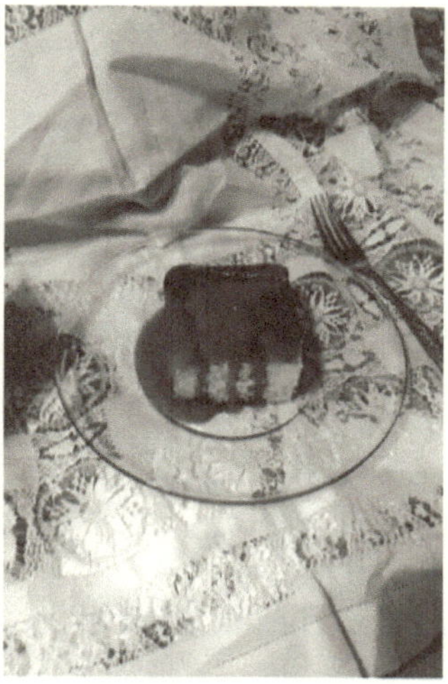

## TWO EGG PUDDIN

1¼ cup sugar

½ cup fresh churned butter, blend together until soft and fluffy

Add two eggs and mix well

2½ cups self-rising flour

1 cup milk

1 teaspoon vanilla or lemon flavoring.

If you have an electric oven, heat it to 350 degrees. We didn't have electricity, just the very hot wood stove. Grease and flour a long baking dish or two 8- or 9-inch round cake pans. Pour in batter, bake until golden brown.

## SAUCE

VANILLA OR CHOCOLATE

2 cups milk

1¼ cups sugar

2 tablespoons flour

Mix sugar and flour and add milk until all is mixed. Place over medium heat, stir constantly and cook until desired thickness. Remove from heat, add one teaspoon vanilla or lemon flavoring.

To make a chocolate sauce use two tablespoons of Hershey's cocoa.

## NEVER FAIL PIE CRUST

1½ cups plain flour

1 teaspoon baking powder

½ teaspoon salt

8 tablespoons (lard) or Crisco

4 tablespoons cold water

Mix all dry ingredients until shortening. Add cold water until a soft roll is formed. Roll out on a floured board until thin. Place in baking dish, removing all bubbles, add your favorite filling and bake at 350 degrees.

# EGG CUSTARD PIES

6 eggs, separated

(keep egg whites for

meringue)

1½ cups sugar

2¼ cups milk

1 teaspoon vanilla

2 heaping tablespoons flour

½ stick of butter (softened) mix well. Pour into prepared crust. Bake at 350 degrees until custard is firm.

## TOPPING

Beat egg whites until stiff in peaks.

Add 6 tablespoons sugar slowly.

¼ teaspoon cream of tarter

½ teaspoon vanilla

Beat egg whites with a wire whisk (electric mixer) until stiff. Add other ingredients. Spread on top of firm pie custard, return to oven, and bake until golden brown. Cool and enjoy.

LEFT-OVER BISCUITS COULD BE THE DESSERT OF THE DAY

## BREAD PUDDING

5 biscuits day-old, 2 cups cubed

2 tablespoons butter or margarine, melted

½ cup moist raisins (use enough hot water to soften)

¼ teaspoon salt

½ cup sugar

2 eggs, beaten

3 cups scalded milk (hot)

¼ teaspoon cinnamon

1 teaspoon vanilla

Combine milk and eggs. Pour over bread cubes and stir in remaining ingredients. Pour into a round eight-inch pan. Place pan into a larger pan deep enough to hold about one inch of water. Bake at 350 degrees for forty-five minutes or until a knife inserted halfway between the center and outside comes out clean.

# BROWN BAG FRIED CHICKEN

Soak chicken in cold salt water overnight in refrigerator or at least one hour before cooking. Rinse (wash) in cold water and let drain.

Use frying chicken cut into pieces or your favorite pieces.

## EGG WASH FOR CHICKEN

3 eggs, beaten

Add milk to desired consistency.

In a large brown paper bag:

2 cups flour (approximately)

Salt, pepper, paprika

Place chicken in egg wash until coated, then put in brown paper bag with other ingredients (about three pieces at a time) Shake until coated with flour mixture.

Cook in a large iron skillet, heated to medium heat, with Crisco or vegetable oil. Brown on one side, turn and brown the other side. Remove and place on paper towel to drain.

# Pictorial Gallery

Josephine (left) and Dorothy Jean (right)

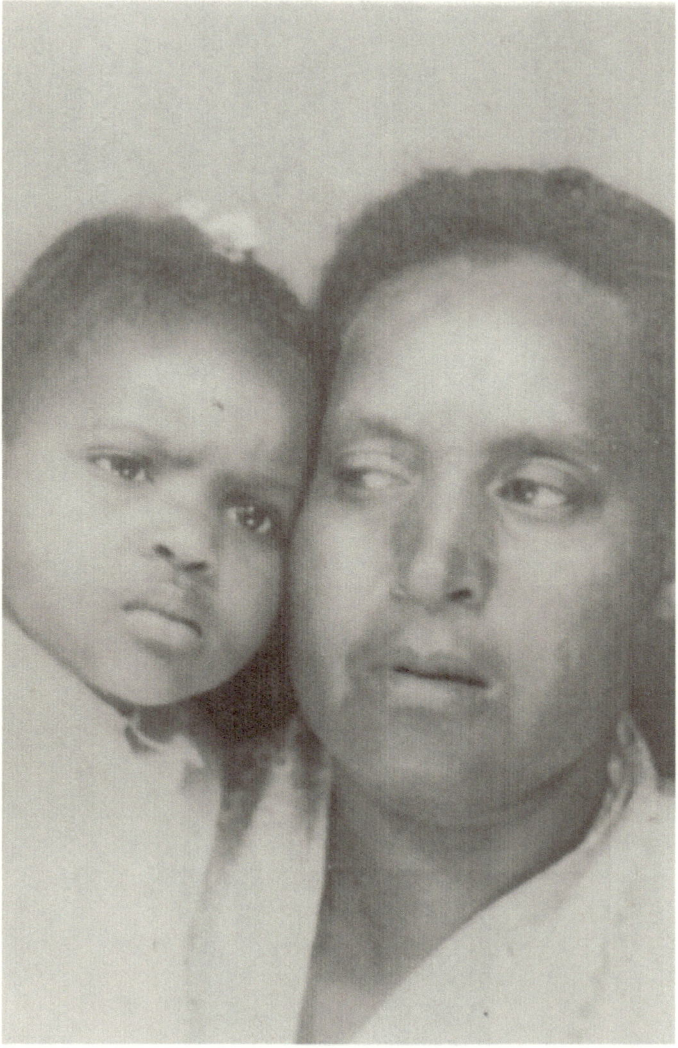

Dorothy Jean (baby girl) Grandmother Hattie Ella Statton

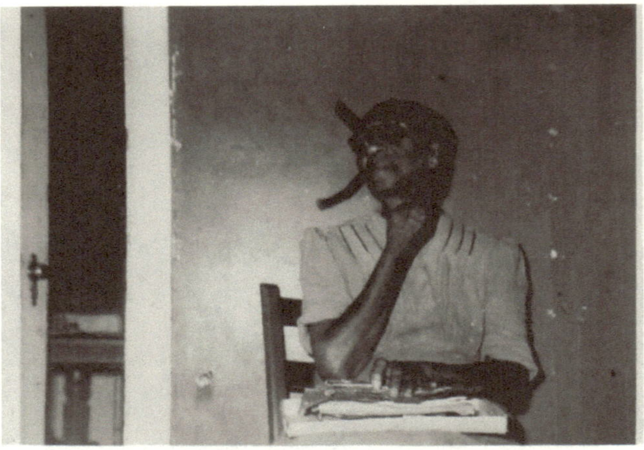

(Mama Della) Idella Allen Woodard

William C (Jack) Abshire and Precious J Woodard Abshire

Josephine (left) and Dorothy (right seated)

Kenneth Wade Tolliver

Janet Elaine Tolliver Hopson

Dorothy Tolliver, Present Day